THUNDER AND LIGHTNING STORMS:
CAUSES AND EFFECTS

by Cynthia O'Brien

12
STORY LIBRARY
MORE TO EXPLORE

www.12StoryLibrary.com

12-Story Library is an imprint of Bookstaves.

Developed and produced for 12-Story Library by Focus Strategic Communications Inc.

Library of Congress Cataloging-in-Publication Data
Names: O'Brien, Cynthia (Cynthia J.), author.
Title: Thunder and lightning storms : causes and effects / by Cynthia O'Brien.
Description: Mankato, Minnesota : 12-Story Library, [2022] | Series: Wild weather |
Includes bibliographical references and index. | Audience: Ages 10–13 | Audience: Grades 4–6
Identifiers: LCCN 2020018056 (print) | LCCN 2020018057 (ebook) | ISBN 9781645821526 (library binding) |
ISBN 9781645821908 (paperback) | ISBN 9781645822257 (pdf)
Subjects: LCSH: Thunder—Juvenile literature. | Lightning—Juvenile literature.
Classification: LCC QC968.2 .027 2022 (print) | LCC QC968.2 (ebook) | DDC 551.55/4—dc23
LC record available at https://lccn.loc.gov/2020018056
LC ebook record available at https://lccn.loc.gov/2020018057

Photographs ©: Leonard Rodriguez/Shutterstock.com, cover, 1; NOAA/CC2.0, 4; egd/Shutterstock.com, 5; No-Plastic/Shutterstock.com, 5; KIDSADA PHOTO/Shutterstock.com, 6; Mihai Simonia/Shutterstock.com, 7; W.carter/CC1.0, 7; Designua/Shutterstock.com, 8; CNN/YouTube.com, 9; Piotr Krzeslak/Shutterstock.com, 9; Andrey Solovev/Shutterstock.com, 10; Peter J. Wilson/Shutterstock.com, 11; Courtesy of National Weather Service, 11; Roman Mikhailiuk/Shutterstock.com, 12; Martin Haas/Shutterstock.com, 12; Rob and Stephanie Levy/CC2.0, 13; David P Howard/CC2.0, 14; Trong Nguyen/Shutterstock.com, 15; John Panella/Alamy, 15; Patrick H/Shutterstock.com, 16; KTVQ News/YouTube.com, 16; Christian Roberts-Olsen/Shutterstock.com, 17; BNO News/YouTube.com, 17; agefotostock/Alamy, 18; Matteo Benegiamo/Shutterstock.com, 18; Famartin/CC4.0, 19; Irina Kozorog/Shutterstock.com, 19; FrameStockFootages/Shutterstock.com, 20; EHStockphoto/Shutterstock.com, 21; Edward Haylan/Shutterstock.com, 21; delcarmat/Shutterstock.com, 22; Natthawat Thailand/Shutterstock.com 23; TTstudio/Shutterstock.com, 23; Gino Santa Maria/Shutterstock.com, 24; Rawpixel.com/Shutterstock.com, 24; MDay Photography/Shutterstock.com, 25; Thavon Phumijan/Shutterstock.com, 25; Roger Coulam/Alamy, 26; Hemmo Vattulainen/CC3.0, 27; Roman Mikhailiuk/Shutterstock.com, 27; ducu59us/Shutterstock.com, 28; Pavel L Photo and Video/Shutterstock.com, 28; Roger Brown Photography/Shutterstock.com, 29; KMH Photovideo/Shutterstock.com, 29

About the Cover
Lightning bolts crackle in Sao Paulo, Brazil.

12

Access free, up-to-date content on this topic plus a full digital version of this book. Scan the QR code on page 31 or use your school's login at 12StoryLibrary.com.

Table of Contents

What's Making All That Racket? 4

What Is Thunder? 6

What Is Lightning? 8

Are Thunderstorms Different from Lightning Storms? 10

What Happens After a Thunderstorm? 12

How Do Thunder and Lightning Storms Affect People? 14

How Do Thunder and Lightning Storms Affect Nature? 16

When Is It Safe? 18

Will a Thunderstorm Happen Today? 20

Is That True? 22

Does Climate Change Affect Thunder and
Lightning Storms? 24

What Are Thundersnow and Fire Clouds? 26

Staying Safe in a Thunder and Lightning Storm 28

Glossary 30

Read More 31

Index 32

About the Author 32

What's Making All That Racket?

Lightning strikes in Texas.

Have you ever heard a loud clap or a rumbling in the sky? Did you see a flash of lightning, too? This is a thunderstorm. These storms can happen anytime of the year. But they are more common in the spring and summer. They also happen often over large cities.

Thunderstorms usually last about 30 minutes to an hour. Severe thunderstorms can last much longer. Most thunderstorms have three stages. During the first part of the storm, a thunder cloud forms. The water inside the cloud becomes heavy. Rain starts to fall.

The cloud grows larger as warm, wet air rises. At the same time, the air below the cloud becomes cooler. After a while, the cooler air stops the warm air from rising. This causes the storm to die out.

THINK ABOUT IT

Why do thunderstorms happen more often in the summer? Why do they occur over big cities?

5

What Is Thunder?

Thunder is the loud noise you hear after you see lightning. You see lightning first because light travels much faster than sound. Lightning causes thunder. The electricity in lightning heats up the air to over 50,000° Fahrenheit (27,760° C). This happens in a fraction of a second. The hot air expands very quickly. Then it shrinks quickly as it cools down. This causes a sharp, loud bang.

5

Time in seconds it takes for sound to travel one mile (1.6 km) through air

- People rarely hear thunder that is more than 10 miles (16 km) away.
- If you hear thunder, go inside right away.
- There is some thunder that people cannot hear. It is called infrasonic sound.

Some thunder sounds are loud booms. Lightning can make this kind of thunder when it strikes the ground. What about rumbling thunder? Thunder makes this noise as the air vibrates. A rumbling thunder means that lightning is far away. If the sound is a loud clap, it means that lightning is close by.

TOWERING STORM CLOUDS

A cumulonimbus cloud is a thundercloud. It can be 60,000 feet (18,000 m) high. These clouds may group together. This can cause severe thunderstorms.

7

What Is Lightning?

HOW LIGHTNING IS FORMED

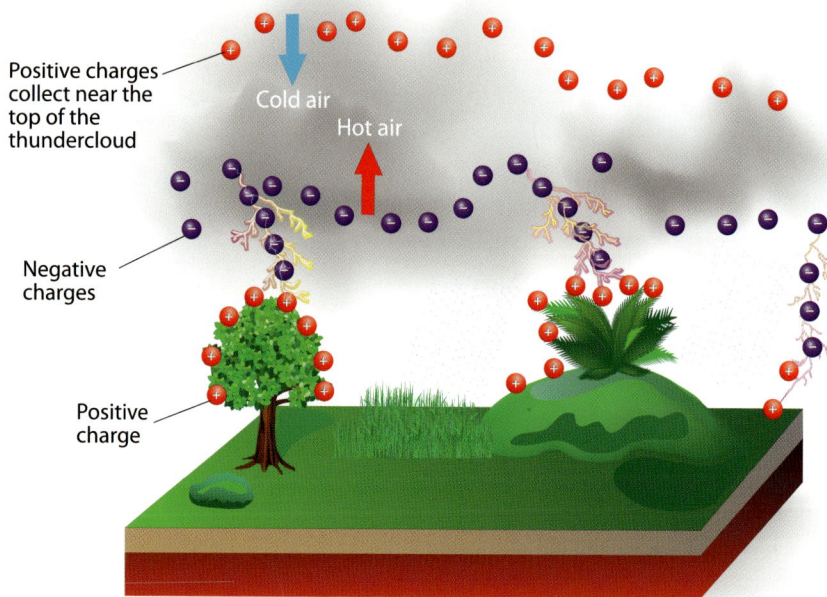

Positive charges collect near the top of the thundercloud

Cold air

Hot air

Negative charges

Positive charge

Lightning is a fierce bolt of electricity in the sky. Most lightning forms inside clouds that are full of water and ice. The air pushes warm water droplets up and turns them into icy particles. Meanwhile, air pushes down other icy particles in the cloud. This causes a lot of reactions inside the cloud. The particles crash against one another and build up an electrical charge. Then, it breaks through the air as a bright, jagged flash of light.

Chile's Calbuco volcano erupted in 2015. It was so big and hot that it caused lightning.

Sometimes, volcanoes cause lightning. When a volcano erupts, it sends hot ash into the air. That makes an ash cloud. The ash particles bump into each other. This builds up electricity inside the ash cloud.

Sparks of lightning hit the ash and heat it up even more. The biggest, hottest eruptions cause lightning.

54,000°

Temperature in Fahrenheit (30,000 °C) of the air from a single lightning stroke

- One flash of lightning can heat the air to five times hotter than the Sun.
- The electrical bolt of lightning is not hot. It merely heats up anything it passes through.
- The energy from one lightning bolt could light a lightbulb for three months.

Are Thunderstorms Different from Lightning Storms?

There are no thunderstorms without lightning. Thunderstorms also come with strong winds. They produce heavy rain or hail (balls of ice). A severe thunderstorm can even cause a tornado.

We see the lightning that strikes from the cloud to the ground. Lightning also travels between clouds. Sometimes we see the sky light up, but no lightning bolt. This is lightning that happens inside a cloud. Dry lightning storms happen without rain. These can cause forest fires.

Lightning strikes can cause forest fires.

Other times, we see lightning but do not hear thunder. If this happens, the thunder may be too far away to hear. Also, the sound waves may bend up and away from the ground. People cannot hear this thunder, either.

1.4 billion
Times per year that lightning strikes around the world

- The most thunderstorms in the United States happen in Florida.
- Thunderstorms can have winds over 100 miles per hour (161 km/h).
- Lightning storms often happen over hot, humid, big cities.

SUPERCELL STORMS

A very severe thunderstorm may be a *supercell*. This is a storm that develops from a number of thunderclouds joining together. It lasts longer than an hour. Winds whip around at great speed. A supercell releases drenching rains. The worst tornadoes form in supercells.

11

What Happens After a Thunderstorm?

A severe thunderstorm leaves behind a lot of wreckage. The wind can rip up plants and trees. It can take down power lines. Lightning strikes can cause fires. Heavy hail can break windows and damage cars. Tornadoes can turn cars upside down and tear down buildings.

Thunderstorm rain causes widespread flooding. A thunderstorm may bring

12

an inch (2.5 cm) of rain. This means 13,577 gallons (51, 395 l) in an average backyard. After a storm, people should never try to drive through a flooded street. It might be deeper than they think.

When you think it is safe to go outside, watch out. There may be power lines down. Do not touch them. Look out for fallen trees, too. If you see anything, report it.

$10 billion
Cost per year of damage to US property and agriculture from thunderstorms

- Lightning causes about half the wildfires in the United States.
- Many injuries come after a storm. Be aware of broken glass and nails.
- Damaged power and gas lines can cause fires, explosions, and electrocution.

THINK ABOUT IT

What should be in your emergency kit so that you are safe and healthy after a storm?

6

How Do Thunder and Lightning Storms Affect People?

Thunder sounds scary, and a thunderstorm brings real dangers. Lightning can be deadly. It can also cause serious injuries, such as burns. Think about what is near you. Stay away from metal fences and wires. The lightning can use these to conduct electricity.

A direct strike may hit people who are working or playing outside. This is not common. If lightning strikes a tree, the electricity travels down. It goes along the ground and strikes people nearby. Lightning bolts branch out. They can strike sideways, too.

Thunderstorms also cause flash floods. These happen when a lot of heavy rain falls in a short time. They are very dangerous because they happen so quickly.

4,000 or more

Number of people worldwide killed by lightning each year

- The number may be much higher. Not all deaths are recorded.
- Each year, 16 million thunderstorms rumble around the world.
- Flash floods from thunderstorms cause more deaths than lightning.

BOLT FROM THE BLUE

Some lightning comes by surprise. This is a *bolt from the blue*. A lightning strike like this can travel as far as 25 miles (40 km) from the cloud.

7

How Do Thunder and Lightning Storms Affect Nature?

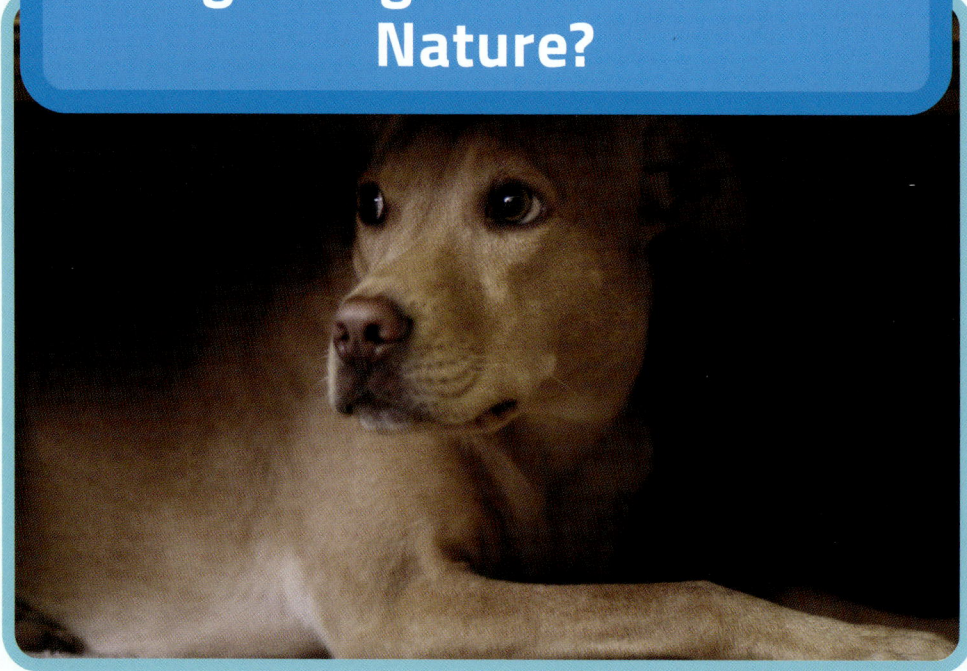

If a dog acts scared and tries to hide, a storm might be coming. Dogs and other animals can hear thunder before people do. Pets are often inside, but wild animals are not so lucky. Thunderstorms bring massive amounts of rain and wind. This can destroy habitats. These storms can also bring hail. In August 2019, hail killed more than 11,000 water birds in Montana.

Hailstones.

Lightning can strike one animal. It can also kill a whole herd. The lightning can jump from one animal to another. Lightning can start fires and burn plants. The heat can destroy a tree completely. It explodes rock and eats away soil.

Thunderstorms cause a lot of destruction. But they are a great way to cool down the Earth. They also release a lot of energy. Thunderstorms help to keep the Earth and the atmosphere in balance.

150
Distance in miles (240 km) that an elephant can sense rain

- Elephants appear to move away in time to escape rain and thunderstorms.
- In Norway, lightning killed 323 reindeer on one occasion.
- The lightning likely struck the ground. A strike can affect animals standing on an area of up to 260 feet (80 m) across.

When Is It Safe?

THANK YOU FOR JOINING US TODAY - I'M AL RUECHEL...

baynews9.com

Bay NEWS 9   Al

11:59
47°

SANYO

Thunderstorms can be dangerous. It's good to know that people are on the lookout for them. The storm prediction center issues a thunderstorm watch. This watch tells people to get ready. It is important to keep listening to alerts for up-to-date information.

National weather offices send out a warning if a thunderstorm is happening. They find out from their radar or from spotters. This is an emergency alert. It may be on a cell phone, the

SEVERE WEATHER ALERT

Thunderstorms can uproot trees.

television, or radio. This alert tells people to find a safe shelter right away.

You can also follow the 30-30 rule. If you see lightning, start counting to 30. If you hear thunder before you get to 30 seconds, go inside. After you hear the last thunder, wait at least 30 minutes to go outside. Check the weather station to be sure.

3
Times per day a severe weather notice is issued

- A severe thunderstorm has winds at least 58 miles per hour (93 km/h).
- It has hail that is one inch (2.5 cm) or larger.
- A watch covers an area of up to 40,000 square miles (100,000 sq km). This is about the size of Virginia or Kentucky.

19

9

Will a Thunderstorm Happen Today?

How do meteorologists know a storm is coming? They look for heat and moisture in the air. Tools such as radar help them to locate the rain or hail. Other radar detects the way the wind is blowing. Satellites show thunderstorms forming. Scientists look at the images to see how severe a storm might be. This is useful for sending out early warnings.

Scientists also make cloud models and test them. They take measurements during

LIGHTNING HOTSPOTS

Where in the world does lightning happen most often? NASA and the Japan Aerospace Exploration Agency wanted an answer. They used a satellite to collect information and make a map. They noticed that more lightning happens close to the equator.

real thunderstorms. Then they study this information. Computers take data from testing stations. They run different tests with the information. This also helps weather scientists to predict how a storm will behave.

900
Number of locations around the world that send out weather balloons

- The locations release two balloons every day.
- A tool on the balloon sends information to the ground every one or two seconds.
- The balloons rise about 20 miles (32 km) into the air and travel for about two hours.

21

Is That True?

In ancient Greek myths, the god Zeus used lightning bolts as weapons. Lightning and thunder appear in the myths of other cultures, too. Today, we know the science behind these weather events.

Still, not everything you hear about thunderstorms is true. You may think that lightning does not strike the same place twice. This is false. Even people can be struck by lightning more than once. If a person is electrocuted by lightning, are they safe to touch? Yes. People do not carry the electric charge.

Another myth advises people lie on the ground in a storm. Lightning actually sends electricity through the earth.

This is ground current. Also, it is not safe just because you do not see a thundercloud or rain. Lightning can strike several miles away from any rain. Going inside is the best thing to do.

THINK ABOUT IT

Lightning electricity can travel through wires and plumbing. What should you do or not do inside the house?

21 to 25

Number of times per year that lightning strikes the Empire State Building

- Lightning strikes Lake Maracaibo, Venezuela, 300 days per year.
- Lightning can strike three miles from the middle of a thunderstorm.
- Ground current can travel 33 feet (10 m) or more.

23

11

Does Climate Change Affect Thunder and Lightning Storms?

Pollution causes heat buildup.

The planet naturally heats and cools over time. Human activities have altered these cycles. For example, pollution in the air traps more heat on Earth.

Tornadoes are difficult to study. But scientists think that there may be more of them. Thunderstorms also seem to happen more often. Climate change could be the reason. However, climate

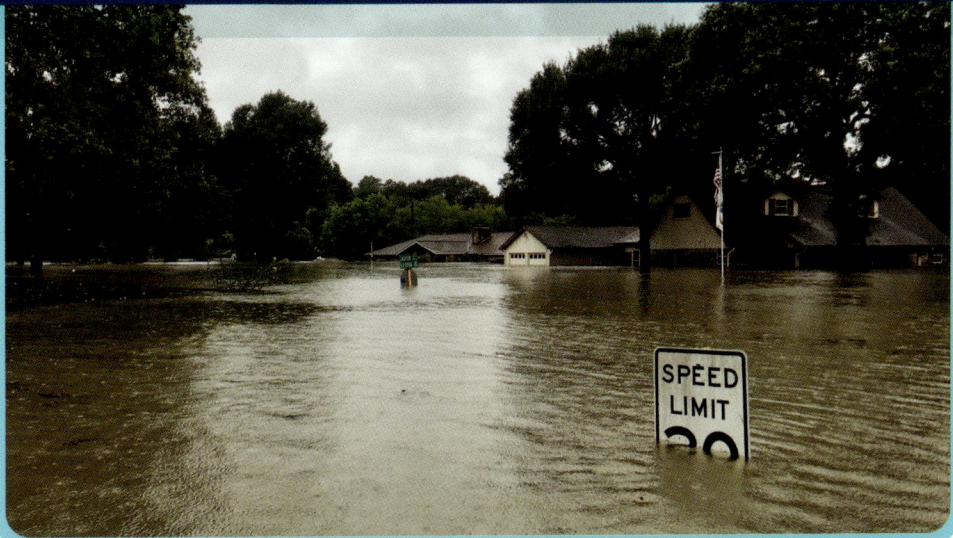

change could mean less difference in temperature between cold and hot places. That might lead to fewer thunderstorms and tornadoes.

Already, rainfall is heavier across the United States. In the future, the warmer Earth might mean more rain during thunderstorms. This would cause more flooding. Lightning already causes many forest fires. But climate change is causing more of these fires in Alaska and northern Canada. As these areas warm up, more lightning fires seem likely.

206

Number of tornadoes in one week in the US in May 2019

- Thunderstorms caused the tornadoes.
- Lightning strikes could increase by 50 percent by the year 2100.
- Warmer air holds more water.

What Are Thundersnow and Fire Clouds?

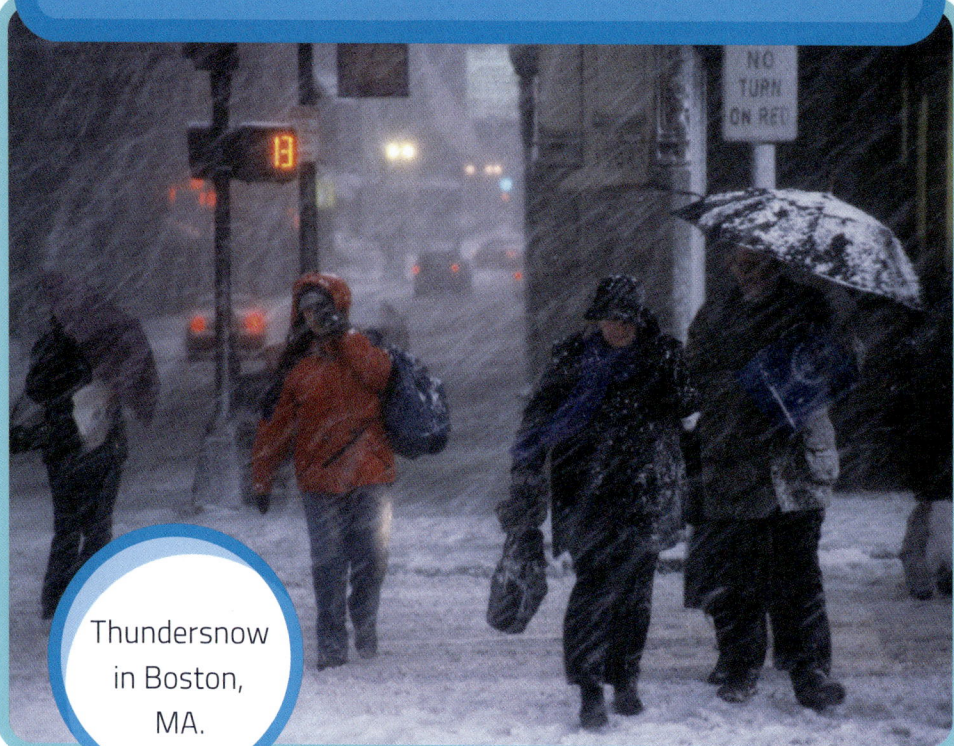

Thundersnow in Boston, MA.

A snowy day does not seem a likely time for a thunderstorm. Thundersnow is a storm that produces snow instead of rain. Thundersnow is not common. It produces less lightning but a lot of snow. These storms are smaller than warm thunderstorms.

They sound different, too. The snow softens the loud thunder claps and rumbling.

Fire clouds form when heat rises quickly from a burning fire below. The smoke condenses when it rises. If the cloud grows very large,

2 to 3

Distance in miles (3 to 5 km) away from lightning when you hear thundersnow

- Less than one percent of snowstorms produce thunder and lightning.
- Less than one percent of lightning's energy is made into sound. The rest is light.
- An average of 25 fire cloud events happen each year in western North America.

it can produce lightning. Lightning strikes can make fire worse. Fire clouds also come with strong winds that can spread fire. All of this can be very dangerous. But fire clouds sometimes produce rain. This helps to put out the fire.

Lightning appears pink when it travels through snow.

WHAT COLOR IS LIGHTNING?

Lightning often appears white, but other times it is different colors. It all depends on particles in the air. If lightning travels through snow, it can appear pink and green. Rain, hail, and dust can turn lightning other colors.

27

Staying Safe in a Thunder and Lightning Storm

SEVERE WEATHER ALERT

- If you hear a roar, go indoors!

- Stay indoors if there are thunderstorm warnings or alerts in your area.

- Don't get electrocuted! Avoid trees, towers, metal fences, and poles.

- Get out of the water. If you are in a boat, get to shore. Head for shelter.

- Move out of areas that could flood, and go to shelter on higher ground.

- Avoid fallen power lines or cables. They have live electricity that could shock you.

- Inside, do not shower or do the dishes. Unplug electrical devices.

- Don't use a telephone land line. This is a major cause of lightning injury.

- Have an emergency kit ready for a severe storm. This should have water, food that won't spoil, and a first aid kit.

Glossary

atmosphere
The gases that surround the Earth and other planets.

condense
Change from a gas or vapor to a liquid.

conduct
Allow electricity to travel from one place to another.

electrocution
To kill by an electric shock or charge.

greenhouse gas
Gas in the Earth's atmosphere that traps the Sun's heat.

meteorologist
A scientist who studies weather and the atmosphere.

satellite
Machine in space that sends information back to Earth.

spotter
Someone who watches and reports patterns, temperatures, and other weather.

tornado
Powerful winds that spin around in a tall column and move quickly. They often form at the base of a thunderstorm.

Read More

Bailer Darice. *Why Does it Thunder and Lightning.* New York, NY: Benchmark, 2010.

Dykstra, Mary. *Climate Change and Extreme Storms.* New York, NY: Lerner, 2019.

Johanson, Paula. *Lightning and Thunder.* Nature's Mysteries. New York, NY: Rosen, 2019.

Stewart, Melissa. *Lightning.* Inside. New York, NY: Sterling, 2011.

Visit 12StoryLibrary.com

Scan the code or use your school's login at **12StoryLibrary.com** for recent updates about this topic and a full digital version of this book. Enjoy free access to:

- Digital ebook
- Breaking news updates
- Live content feeds
- Videos, interactive maps, and graphics
- Additional web resources

Note to educators: Visit 12StoryLibrary.com/register to sign up for free premium website access. Enjoy live content plus a full digital version of every 12-Story Library book you own for every student at your school.

Index

animals, 16–17

benefits, 17

climate change, 24–25
clouds, 7
color of lightning, 27
cost of thunderstorms, 13

damage, 12–13
dangers, 14–15
deaths, 15
definition of lightning, 8–9
definition of thunder, 6–7
distance lightning
    travels, 15
dry lightning storms, 10

elephants, 17
Empire State Building, 23

fire clouds, 26–27
flooding, 12–13

hotspots, 21

injuries, 14–15

meteorologists, 20–21
myths, 22–23

number of lightning
    strikes, 11

predicting thunderstorms,
    18–19, 20–21

safety, 19, 22–23, 28–29
satellites, 20
size of lightning, 5
sounds of thunder, 7
speed of sound, 7
stages, 4–5
supercell storms, 11

temperature, 9
thundersnow, 26–27
timing of thunderstorms, 5
tornadoes, 25

volcanoes, 9

weather alerts, 18–19
weather balloons, 21

About the Author

Cynthia O'Brien has lived and worked in Canada and England. She has written many books for children and young readers, including books about science, cultures, and explorers.

READ MORE FROM 12-STORY LIBRARY

Every 12-Story Library Book is available in many formats. For more information, visit **12StoryLibrary.com**